RAILW
WORKSHOPS

Tim Bryan

SHIRE PUBLICATIONS

Published in Great Britain in 2012 by Shire Publications Ltd, Midland House, West Way, Botley, Oxford OX2 0PH, United Kingdom.

44-02 23rd Street, Suite 219, Long Island City, NY 11101, USA.

E-mail: shire@shirebooks.co.uk www.shirebooks.co.uk

A CIP catalogue record for this book is available from the British Library.

Shire Library no. 707. ISBN-13: 978 0 74781 201 2

Tim Bryan has asserted his right under the Copyright, Designs and Patents Act, 1988, to be identified as the author of this book.

Designed by Tony Truscott Designs, Sussex, UK and typeset in Perpetua and Gill Sans.

Printed in China through Worldprint Ltd.

12 13 14 15 16 10 9 8 7 6 5 4 3 2 1

COVER IMAGE

A painting by Terence Cuneo of the 'A' Erecting Shop at Swindon Works, for a 1957 British Railways poster titled 'Progress'. Instead of a Swindon-built steam locomotive, the poster featured a new Western Region Diesel-Hydraulic 'Warship' class engine.

TITLE PAGE IMAGE

Repairing a massive Maybach diesel locomotive engine at Swindon in the 1970s. This building had formerly been the iron foundry.

CONTENTS PAGE IMAGE

A line of locomotives outside Crewe Works in the 1930s. The nearest is LMS 4-6-2 no. 4621 *Duchess of Atholl*.

ACKNOWLEDGEMENTS

I am very grateful for the assistance of Elaine Arthurs, Collections Officer at STEAM: Museum of the Great Western Railway in Swindon for her help in the preparation of this book, particularly in making accessible the collections of the late Alan Peck. I would also like to thank the volunteers at the STEAM library for their help during my research. Thanks are also due to staff at Derby College and Derby Tourist Information Centre and to Laurence Waters at the Great Western Trust and John Whalley.

As always, thanks are also due to my wife Ann for her encouragement and support, and for reading and editing the manuscript.

I am grateful for the following for permission to reproduce illustrations in the book:

Derby College, page 52; Great Western Trust Didcot, pages 30 (bottom), and 33 (bottom); Keith Long, page 10, and 35 (bottom); R.P. Marks, page 16; by courtesy of the Mitchell Library, Glasgow City Council, pages 7, 19, 24 (top) and 43 (bottom); Science & Society Library (front cover); John Whalley, page 48 (top). Pictures from the author's collection are reproduced on pages 4, 15, 17 (top right), 20, 23, 24 (bottom), 28 (top), 45 (bottom) 53, 55; all other photographs are from the STEAM collection.

CONTENTS

SWINDON WORKS

GWR

PRICE 2

INTRODUCTION

EVEN A CASUAL VISITOR to one of the great railway works such as those at Crewe, Doncaster, Glasgow, Manchester or Swindon in the early twentieth century would have been impressed by the sheer scale of the industry. Behind the great walls of these factories, workshops resounded to the noise of thousands of men building locomotives and rolling stock both for the railways of Britain and for export to all parts of the world. But now, nearly two hundred years after Robert Stephenson & Company set up the first proper factory to manufacture railway engines in Newcastle, the British railway manufacturing industry is a mere shadow of the enormous engineering business that, at its height, was one of the country's biggest employers and built over two thousand locomotives every year for railways at home and abroad.

The industry began in the cradle of railways, the north-east of England, with small engineering concerns building engines for early railways such as the Stockton & Darlington and the Liverpool & Manchester. Once railways had become firmly established, the industry developed dramatically as new companies were set up to supply railways with the locomotives and rolling stock they needed to operate. With the expansion of the network, railways soon found it necessary to build their own engineering workshops to construct and maintain not only locomotives, but also carriages, wagons and other equipment, beginning a proud tradition of railway development that was to last more than 150 years.

Places such as Crewe, Derby, Doncaster, Eastleigh and Swindon soon became famous as 'railway towns', with new communities growing up alongside the extensive railway workshops, housing thousands of men and their families. The scale of these operations was enormous, with the works running as self-contained establishments, having their own foundries, machine shops, erecting shops, woodworking and carriage body shops. Supporting all this were huge office and design complexes, electric and hydraulic power stations, and other associated activities.

As well as the works operated by the main-line railways, a large and important private industry continued to flourish despite enduring difficult

Opposite: The front cover of a guide to Swindon Works issued by the GWR in the 1930s.

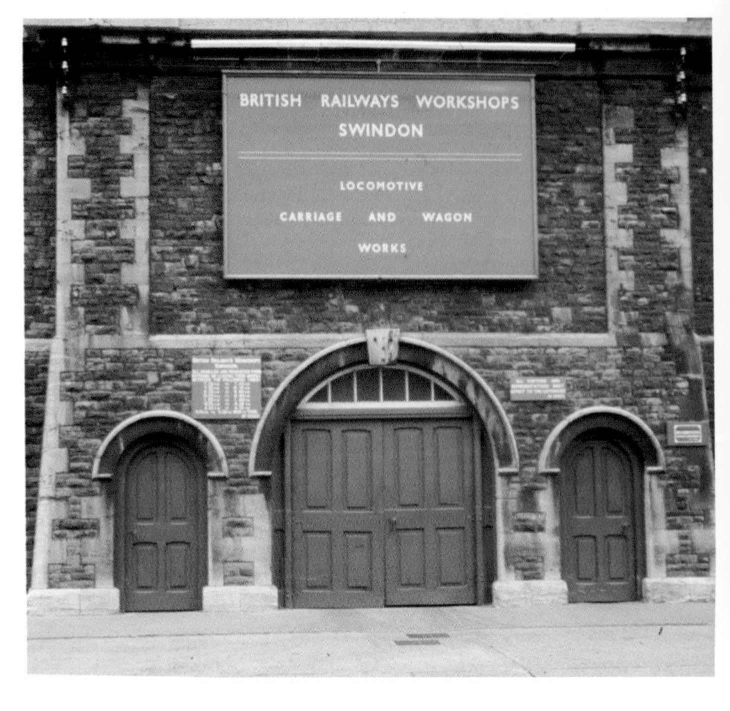

The doors are now firmly closed but behind them thousands of men and women built locomotives, carriages and wagons for the Great Western Railway and British Rail at Swindon.

times in the years between the two world wars, with famous firms such as Beyer Peacock, North British, Robert Stephenson and Vulcan Foundry not only building engines for domestic use, but exporting British railway locomotives and rolling stock all over the world.

A view of the North Eastern Railway works at Shildon taken from the window of the works manager's office in 1946.

A 2-8-2 locomotive for East African Railways is suspended above the erecting shop of the North British Locomotive Company's Glasgow works in 1952.

After 1948 more than 47,000 people were still employed in workshops run by the nationalised British Railways (BR), with a further thirteen thouand working for private firms. Despite a brief renaissance in the post-war period, the gradual decline of the railway manufacturing industry was inextricably linked to the dramatic changes experienced by the railways since 1945. The end of steam on British Railways, its replacement by diesel and electric motive power following the BR Modernisation Plan, and the eventual privatisation of British Rail have had a profound effect on the railway-owned workshops. Growing competition from railway builders in Europe, Asia and the United States eventually led to the end of many private builders.

While railway passenger numbers are at their highest level for many years, Britain's railways are now supplied with new locomotives and rolling stock largely by overseas manufacturers. Derby's Litchurch Lane factory is the last remnant of an industry that dominated the world, although a number

The spartan canteen facilities provided for workmen at the Inverurie locomotive works of the LNER, pictured in 1946.

of other old workshops survive as maintenance facilities. In other railway towns, workshops have been demolished or given new uses. Derby Locomotive Works now houses a further education college, and at Swindon shoppers now walk where workers once built locomotive boilers for the Great Western Railway (GWR).

The construction of wagon stock was one of the more mundane tasks undertaken by railway workshops. Note that the two workmen operating the noisy hydraulic rivet guns are not wearing any ear production.

Left: A diesel locomotive bogie under repair at the Brush locomotive works in 1995.

Below: The main erecting shop of Darlington Works in September 1959. By this time the construction of new steam locomotives had finished, and so the engines pictured were in the works for overhaul or repair. Gresley V2 no. 60932 is nearest the camera.

68736
XP

A BRIEF HISTORY

THE FIRST successful steam locomotives were constructed at the collieries and industrial complexes that were at the heart of the railway revolution in the nineteenth century. Workshops at Coalbrookdale and Penydarren built engines designed by Richard Trevithick in 1802 and 1804, and George Stephenson's first locomotive, *Blücher*, was built at Killingworth Colliery in 1814. Other engines, including the 1813 *Puffing Billy*, were manufactured at Wylam Colliery, and engines for the Middleton Colliery were built nearby in Leeds.

All of Stephenson's earliest engines were built for use at Killingworth, but in 1817 he was commissioned to supply a locomotive for the Kilmarnock & Troon Railway, and this was built in a Newcastle ironworks. Realising the potential of the growing interest in railway schemes, George and his son Robert set up what became the world's first railway factory, the Forth Street Works in Newcastle, in 1823. It was here that Robert Stephenson and Company's first engine, *Locomotion*, was completed for the Stockton & Darlington Railway in 1825. By the time the company built *Rocket* in 1829, railway development was in full swing and, following the opening of the Liverpool & Manchester Railway the following year, Robert Stephenson & Company were supplying engines to many of the major new railways in Britain and abroad.

Despite the Stephensons' dominant position, competitors soon appeared to challenge them, many of which were existing engineering companies that had moved into the railway business.

As the network grew, railways began to struggle with the task of maintaining their locomotives, and it soon became apparent that facilities to overhaul and repair engines and rolling stock were required. Within twenty-five years of the opening of the Stockton & Darlington Railway, most railways had established their own workshops. Although the Stockton & Darlington had opened its own locomotive works as early as 1826, most of the works that later became well-known to railway enthusiasts were established after 1840. Some, such as Crewe and Swindon, were built on greenfield sites

Opposite: The smokebox of 0-6-0 68736 at the Darlington North Road Scrapyard, 26 January 1964.

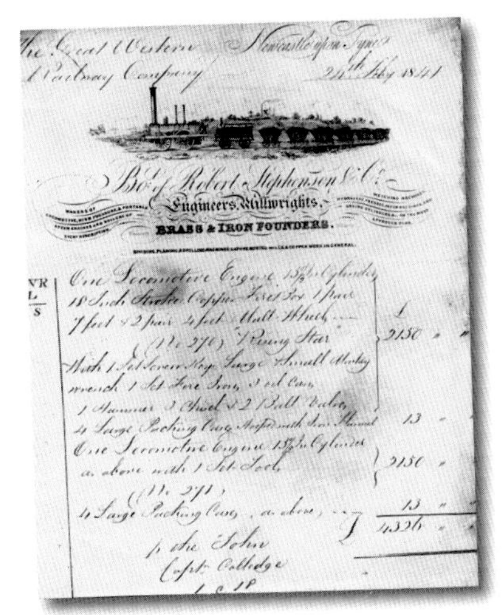
The Great Western Railway Company — Newcastle upon Tyne, 24th Feby 1841

Bo't of Robert Stephenson & Co.
Engineers, Millwrights,
BRASS & IRON FOUNDERS.

	£
One Locomotive Engine 15½ in Cylinder 18 Inch Stroke Copper Fire Box 1 pair 7 feet & 2 pair 4 feet Wheels (No 270) "Rising Star"	2150
With 1 Set Screw Keys Large & Small Monkey wrench 1 Set Fire Irons, 3 oil Cans, 1 Hammer 3 Chisel & 2 Ball Valves, 4 Large Packing Cases	13
One Locomotive Engine 15½ in Cylinder as above with 1 Set Tools (No 271)	2150
4 Large Packing Cases, as above	13
	4326

The original invoice from Robert Stephenson & Company for two locomotives, *Rising Star* and *North Star*, which cost the GWR £4,326 in February 1841.

where there was no local tradition of engineering, while others, such as Ashford, Brighton, Derby, Darlington and Doncaster, were set up close to existing towns. A third group comprised works established in the heart of crowded urban areas, such as the London & South Western's Nine Elms factory in London, and the North Eastern Railway plant at Gateshead. From modest beginnings, most factories expanded their production and workforces during the nineteenth century so that by 1899 the total annual production of locomotives in company workshops was over six hundred, a figure never exceeded, even after the First World War.

Although railways would eventually build many of their own carriages and wagons, many of the carriages used on early lines were constructed by coachbuilding companies with existing skills and traditions. A typical example was the Birmingham company that eventually became known as Metropolitan Railway Carriage & Wagon Ltd. It was set up in 1845 by the London stagecoach builder Joseph Wright, who had already built carriages for the London & Birmingham and London & Southampton railways as early as 1838. By 1900 there were several established companies building rolling stock for both British and overseas railways, and providing wagons for collieries and industrial concerns.

Many of the main-line railways extended their own workshops to include carriage and wagon repair and construction facilities or built new purpose-built carriage and wagon works. At Derby, the Midland Railway's Litchurch Lane carriage works was opened in 1876, thirty-six years after a locomotive

Photographed at the LNER works at Faverdale, Darlington, in 1925, at the celebrations to mark the centenary of the opening of the Stockton & Darlington Railway, the locomotive is the replica *North Star*, built by GWR apprentices at Swindon. The original had been scrapped by the company in 1906.

works had been established in the town. The GWR did not build its own purpose-built carriage and wagon works until 1869, when it concentrated all work at Swindon, following an unsuccessful attempt to build a carriage works at Oxford. Wolverton Works was initially intended to build and repair locomotives and rolling stock for the London & Birmingham Railway but became a carriage and wagon plant when locomotive building ceased there in 1877. The North Eastern built a works at York in 1865 to repair wagons and horseboxes, followed in 1884 by a new carriage works, concentrating all carriage production there.

With most major British railways building their own stock, the independent locomotive manufacturers turned their attention to the production of locomotives for industry and export, and for small domestic railways. Many famous builders turned out thousands of small tank engines for use in factories, mines, power stations and quarries, and the products of companies such as Bagnall, Andrew Barclay, Hudswell Clarke, Hunslet, Manning Wardle, Peckett and Kerr Stuart still survive in use on preserved railways today.

In 1828 Stephenson's had supplied the first British-built locomotive to the United States, and in the following decade built engines used in Russia, Belgium, France, Germany, Austria and Canada, with other manufacturers supplying motive power for other European states. As indigenous railway industries grew up in these countries, manufacturers turned their attention elsewhere, most successfully to the British Empire. From 1852, when Vulcan Foundry built its first steam locomotive for India, output grew steadily until it peaked in 1907.

As well as building engines for railways abroad, manufacturers also still produced locomotives for British railways that did not have the resources to construct their own. These included many smaller South Wales lines such as the Cardiff and Rhymney railways, and less well-funded enterprises such as the Hull & Barnsley Railway and the Midland & South Western Junction line.

By the First World War, most major railway works had been modernised with the introduction of new machine tools, powerful cranes, and the widespread use of electricity and hydraulic power throughout. There was little opportunity to improve matters further during the war, since railway works found themselves undertaking increasing amounts of munitions production and other work for the war effort. By 1918 the combination of labour shortages caused by conscription and a large backlog of both maintenance and new locomotive building left the railways in a far poorer state than they had been four years previously.

Between the world wars the railway manufacturing industry was hit hard both by the slump that crippled the world economy and by the impact of foreign competition. Although some builders benefited from large orders in

GWR pannier tank no. 5700, one of 250 built by private contractors in 1929. This engine was constructed by the North British Locomotive Company in Glasgow.

the aftermath of the war, supplying engines to countries whose railway infrastructure had been badly damaged in the conflict, business was generally hard to find. In the early 1930s there were years when companies such as Robert Stephenson and North British did not build a single steam locomotive; although those firms survived, others went out of business or were absorbed by competitors.

The utilitarian lines of the cab of an 'Austerity' 2-8-0 built during the Second World War.

The availability of loans from government to the 'Big Four' railways, supplied in an effort to combat unemployment following the Wall Street Crash, did provide some relief to manufacturers. The North British Locomotive Company built engines for the London Midland & Scottish Railway (LMS), the London & North Eastern Railway (LNER) and the Southern Railway (SR), while the GWR placed an order in 1929 for 250 tank engines with six different companies. Nevertheless, the main-line railways still built more engines than their private competitors between the wars, in the process rationalising their operations to larger, more efficient factories, since a consequence of the 1923 'Grouping' was that the Big Four companies had inherited a number of smaller old-fashioned workshops from the railways they had absorbed.

With the outbreak of the Second World War, both railway workshops and independent manufacturers were pressed into action to help the war effort. In 1942, orders were placed with the North British Locomotive Company and Vulcan Foundry for War Department 'Austerity' locomotives for use in continental Europe. In two years, both companies built more than a thousand locomotives in contracts worth over £8 million. Other companies, such as Hunslet and Robert Stephenson, also built smaller tank engines for the War Department. In the workshops of the main-line railways, workers found themselves building equipment for the war effort, including aircraft, tanks and landing craft, as well producing millions of tons of munitions.

After 1948 independent locomotive builders capitalised on a boom in business, building engines for railways in Australia, Africa, India and South America. After 1950 shortages of steel, labour problems and increasing competition from Europe, Japan and the United States began to effect exports, as did the decline of steam and its replacement by diesel and electric traction. By the end of the decade many major manufacturers, such as Beyer Peacock, North British, Robert Stephenson and Vulcan Foundry, had all built their last steam locomotives.

With the nationalisation of British railways in 1948, workshops formerly operated by each of the Big Four companies came under the control of BR regions. There was a huge backlog of maintenance for the 47,000 staff employed in the works to make up, a process made slower by shortages of

A worker wheels his bicycle across the cathedral-like space of the 'A' Erecting Shop at Swindon not long before the closure of the works in 1986.

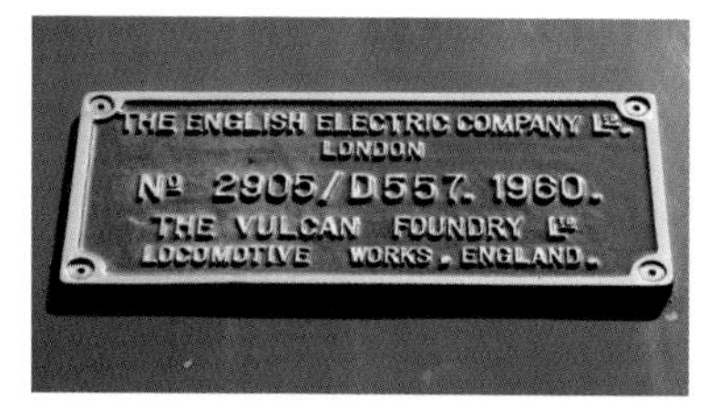

The works plate of a BR Eastern Region 'Deltic' diesel locomotive, built in 1960.

raw materials and capital investment. Works such as Crewe, Doncaster and Swindon built a number of new engines based on pre-war designs, but production of these ended when construction of the new BR Standard steam locomotives began in 1951, with each works nominally building a particular class, an arrangement that was later abandoned.

The adoption of the British Railways Modernisation Plan in 1955 effectively dealt a death blow to many railway works. The plan advocated the end of steam power on BR and the rapid adoption of diesel and electric traction. The last steam locomotive built for British Railways, the 2-10-0 *Evening Star*, was completed at Swindon in 1960, and, although some works built new diesel locomotives and multiple units, most work was transferred to companies such as English Electric, Metrovick, Vulcan Foundry and Robert Stephenson. In 1962 responsibility for railway workshops was removed from the BR regions with the setting up of a separate workshops division. Emphasis shifted from new construction to repair and maintenance, with the consequence that famous works such as Brighton, Cowlairs, Gorton and Stratford were all closed by 1968, along with more than a dozen carriage and wagon establishments.

In 1970 there was a further reorganisation, with workshops transferred to a new company, British Railways Engineering Limited (BREL), which had a remit to look for work inside and outside the BR network. This led, for example, to Swindon Works building diesel shunters for Kenya, and overhauling London Underground stock in later years. Even this new-found independence could not save Swindon, and it closed in 1986. The following year four works, Doncaster, Eastleigh, Glasgow and Wolverton, were moved into a new organisation called BR Maintenance Limited, while Crewe, Derby Locomotive Works, Derby Litchurch Lane Carriage Works and York remained as part of a new BREL company, concentrating on new work and heavy repairs. In 1988 the company was sold into the private sector, becoming part of a consortium including Trafalgar House and Asea Brown Boveri (ABB), which subsequently

Although this image of the trimming shop at Eastleigh Works dates from 1973, the working methods used had remained the same for very many years.

Far left: The first power car of the Advanced Passenger Train (APT-E) takes shape in the Advanced Projects Laboratory of Derby Works in 1971.

Above: This photograph of an 0-8-0 diesel shunter built at Swindon Works for a railway in Kenya featured on a Christmas card sent by the works manager, Harry Roberts, in 1979.

took complete control of the operation. The company became part of the Adtranz group in 1996, and was taken over in turn by the Canadian company Bombardier Transportation in 2001.

Bombardier's plant at Derby, situated in the old Litchurch Lane carriage and wagon works, remains as the only major specialist train manufacturer in Britain. Despite the award of a contract for new Thameslink rolling stock to a European competitor in June 2011, Derby works still survives as the only substantial British railway works with the capacity to build railway locomotives and rolling stock on any significant scale.

Left: A line of ex-BR Class 20 diesels then owned by Rail Freight Services stands outside Doncaster Works in 1991 awaiting overhaul.

1922

INDEPENDENT RAILWAY BUILDERS

ALTHOUGH many British railway companies established their own workshops to build and maintain locomotives and rolling stock, there were also several independent manufacturers who built locomotives and rolling stock for both domestic and overseas markets. The history of the industry is complicated, especially as some manufacturers changed names or merged; writing in the 1970s, the railway historian James Lowe calculated that there had been more than 350 companies, including the railways themselves, involved in the building of steam locomotives from the early part of the nineteenth century until the end of steam.

The two industries existed side by side for more than 150 years but, while railway companies bought locomotives and rolling stock from independent manufacturers, they were prevented from building locomotives other than for their own use, as a result of legal action taken by private locomotive builders in 1875, when the London & North Western Railway had begun to manufacture engines for the Lancashire & Yorkshire. Nearly a century later this arrangement was extended to include the construction of diesel and electric locomotives.

When railways were in their infancy, because the first companies operating train services did not yet have the capability or resources to manufacture locomotives and rolling stock, they relied on a network of builders to supply them with motive power. By 1855 the company set up by George and Robert Stephenson in Newcastle in 1823 had already built more than a thousand engines. These included not only *Rocket* but also the 2-2-0 'Planet' and 2-2-2 'Patentee' types. Many of the latter design were supplied to railways all over Europe and to the United States. In 1839 two engines originally ordered for the New Orleans Railway in the United States were instead modified for use on Brunel's broad-gauge GWR and renamed *North Star* and *Morning Star* respectively.

Stephenson's supplied engines to railways all over the world, to countries as diverse as Australia, Egypt, the Netherlands, India and Turkey. In 1882 the company exported its first engine to China, and the volume of orders

Opposite: A 4-6-2 locomotive, built by the North British Locomotive Company in Glasgow, is loaded on board ship ready for its journey to India in 1952.

This Taff Vale Railway 'D' Class saddle tank engine no. 257 was originally built by the Sharp Stewart Company in 1865. It was rebuilt in 1882 and 1893, taken out of use in 1897, and finally condemned in June 1906.

continued to grow to the extent that by 1900 more than three thousand engines had been built in Newcastle. The following year, to cope with this demand, the company relocated to a new 54-acre site in Darlington. Much of the work it did before the First World War was for export, particularly to parts of the British Empire, such as India and South Africa.

During the First World War, like many railway-owned workshops, Stephenson's carried out a good deal of munitions work, but they also built engines for the War Department. Orders from railways at home or abroad were harder to come by between the two world wars, and Stephenson's acquired the smaller Hawthorn Leslie company in 1937, and both Kitson and Manning Wardle the following year. These acquisitions enabled the enlarged business to build large numbers of smaller tank locomotives for industrial railways for the home market before and after the Second World War.

ABMTM
ASSOCIATED BRITISH MACHINE TOOL MAKERS
17 Grosvenor Gardens, LIMITED London, S.W.1

'BUTLER' 10in. Stroke PUNCHER SLOTTER with built-in circular table, patent slotter drive, pump lubrication, four speed changes, portable pendent control switch. Built for heavy work and easy operation.
Range of sizes 10in. to 48in.
Send for Catalogue No. 4.

'BUTLER' PLANERS, SHAPERS and SLOTTERS.
TYPICAL RAILWAY MACHINES SUPPLIED TO SWINDON WORKS. ILLUSTRATED BY COURTESY OF MR. C. B. COLLETT (Chief Mech. Eng.).

'BUTLER' AXLE BOX PLANER for AXLEBOXES, CROSS-HEADS and CONNECTING ROD BRASSES. Special features include: Dual Control, Clutch Drive, Trunnion mounted Link, Automatic Lubrication, etc.
Catalogue No. 1C will be sent on application.

THE BUTLER MACHINE TOOL CO. LTD., VICTORIA IRONWORKS, HALIFAX

As well as independent locomotive builders, the railway workshops industry also supported a large tool-making and machinery industry, as this 1935 advertisement shows.

Away from the north-east, another early locomotive builder was the Vulcan Foundry, established at Newton-le-Willows near Warrington in 1830 following the opening of the Liverpool & Manchester Railway. The original operation was a partnership between Charles Tayleur, a local businessman and director of the railway, and Robert Stephenson, a relationship that lasted less than two years. The company built engines for railways all over the world, completing its first for India in 1852, one of many built for that country over the next century. The company was also notable for building narrow-gauge 'Fairlie'-type articulated locomotives for use in New Zealand and South America and for the now preserved Ffestiniog Railway. It continued to supply locomotives for both the home and overseas market, with the majority of orders coming from India and Argentina.

A 1933 advertisement boasted that the company had built over two thousand locomotives for India at an average rate of one per fortnight. This work continued until after the Second World War, when orders were lost to builders in mainland Europe, Japan and the United States, and a new industry within India itself. With the demise of steam imminent, the factory was re-equipped for diesel and electric production, and the company was absorbed by English Electric in 1955, by which time workers at the factory had built 6,210 engines in total.

Another famous British locomotive builder was the Manchester company of Beyer Peacock, founded in 1854. Its first major order, for six standard-gauge 2-2-2 locomotives for the Great Western, was completed a year later. Although the company built considerable numbers of engines for railways abroad, it also supplied many British lines with locomotives. By the early part of the twentieth century it had completed more than 1,600 engines for the home market, including locomotives for the Great Central, London & South Western, London & North Western and Metropolitan railways. Beyer Peacock became particularly well-known for building large numbers of Garratt articulated engines, a design characterised by the fact that the boiler was carried on a central frame, supported at either end by separate power units. Although the first two built were 2-foot-gauge engines for a railway in Tasmania, the company produced Garratt designs for all gauges, with the largest being a number of 262-ton 4-8-2-2-8-4 locomotives for Russia in 1932. As well as constructing numerous engines of this type for export, Beyer Peacock also built Garratt locomotives for both the LMS and the LNER in the 1930s. By the 1960s trading conditions were very difficult for the company and, despite retooling its works to build diesels, it went out of business in 1966.

The works plate of British Railways diesel hydraulic locomotive no. D7029, built by Beyer Peacock in 1962

The elegant lines of Lancashire & Yorkshire Railway 4-4-0 no. 995, built for the company by Beyer Peacock in December 1888. The engine was withdrawn in 1933.

A group of railway dignitaries, including former GWR Chief Mechanical Engineer F. W. Hawksworth, stands in front of a massive Beyer-Garratt locomotive in 1952.

Away from the main railway network, many docks, factories, mines, power stations and other industrial concerns relied on having smaller steam locomotives to operate successfully. By 1900 many smaller private locomotive builders existed to satisfy this demand. One such, the Hunslet Engine Company of Leeds, was established in 1864 and built numerous four- and six-coupled standard-gauge industrial tank locomotives, as well as many narrow-gauge engines for slate quarries in North Wales. Two other prominent builders of industrial locomotives were the Scottish companies

Andrew Barclay and Dübs; Barclay built hundreds of 0-4-0 saddle-tank engines for many collieries and steelworks, and also became known for producing 'fireless' locomotives used in munitions depots and factories where fire was a particular hazard.

The Dübs company, which began life in 1865, constructed well over 4,400 locomotives at its Polmadie works in Glasgow, exporting many to China, Cuba, India, Russia, New Zealand and Spain. The enormous demand for engines of this type meant that other companies such as Bagnall, Fox Walker, Hudswell Clarke, Kerr Stuart, Manning Wardle and Peckett all competed for business. In the 1920s the difficult economic conditions endured by the railway industry led to a number of these smaller companies going out of business or being amalgamated with larger competitors.

A significant amalgamation had already taken place at the beginning of the twentieth century when the North British Locomotive Company (NBL) was created following the combination of three Glasgow firms, Dübs, Neilson Reid and Sharp Stewart. The new company became the largest locomotive builder outside the United States and produced many engines for both British and overseas railways. After the First World War NBL built engines for all

The narrow-gauge 0-4-2 *Edward Thomas*, now preserved on the Talyllyn Railway, was originally built for another Welsh slate line, the Corris Railway, by Kerr Stuart of Stoke-on-Trent in 1921.

The front cover of this NBL publication includes illustrations of both an Egyptian sphinx and a pyramid and Constantinople, emphasising its overseas railway export business.

the Big Four companies, including thirty 'King Arthur' express locomotives for the Southern, fifty 'Royal Scot' 4-6-0s for the LMS, and the high-pressure 4-6-0 engine *Fury* for the same company. At the height of the depression, North British also constructed one hundred GWR pannier-tank locomotives and undertook other work for both the LMS and the LNER. By the Second World War, North British had built more than eight thousand engines, a figure that rose to over eleven thousand by the time the company went out of business in 1962. Although NBL was able to secure export orders for locomotives for railways in South Africa, after 1945 increasing competition from the United States and lack of capital to invest heavily in diesel production led to its eventual liquidation.

After the Second World War, despite an initial boom in exports for companies such as Beyer Peacock, NBL and Vulcan Foundry, many famous names began to disappear. The advent of the BR Modernisation Plan and the prospect of private builders being able to produce new diesel and electric locomotives should have been an opportunity for survival, but even some of the best-established firms were unable to continue. The English Electric Company was one operation that did capitalise; the company had taken over both Robert Stephenson and Hawthorn and Vulcan Foundry in 1955, and Metropolitan-Cammell, NBL and Beyer Peacock were all absorbed into the larger group. English Electric was itself taken over by the American company General Electric (GEC) in 1971. GEC had been building diesel and electric locomotives since the 1930s and had considerable expertise both in railways and other areas of engineering. The British part of the new company was sold as part of a further merger in 1989, becoming GEC Alsthom.

An advertisement from the 1930s for one of the biggest independent rolling-stock manufacturers, Metro-Cammell.

The Brush company, which had begun life in 1899, had specialised in building trams and electric locomotives early in its history and was prominent in constructing diesel and electric stock for British Railways in the 1960s and 1970s, most notably the ubiquitous Class 47 diesel designs. In the 1990s the company played a key role in supporting the Channel Tunnel project, constructing 'Le Shuttle' engines at their Loughborough works, as well as new Class 92 locomotives for British Railways. In recent years the company has concentrated largely on locomotive and rolling-stock overhaul and refurbishment rather than new construction, and in February 2011 was purchased by the American Wabtec Corporation.

Above: The Brush works at Loughborough in 1995, with a refurbished Class 20 locomotive suspended above a new Class 92 electric locomotive, *Victor Hugo*, one of a series built for use on both sides of the Channel Tunnel. Brush worked as part of a consortium with ABB Traction.

Left: Independent builders also built rolling stock for overseas use. This carriage body being loaded at Cardiff was destined for Madras, on the Southern Railway of India.

BUILDING LOCOMOTIVES

TO GAIN AN INSIGHT into the operation of a locomotive works in the age of steam, the public would have been able to visit a factory such as Doncaster or Swindon on one of the official tours run by railway companies. Modern health and safety rules would probably make such visits difficult today, even if the works were still operational, so pictures, text, film footage and the reminiscences of workers are the only reminders we now have of the noise, bustle and activity of the workshops run by main-line railways or independent builders.

Many locomotive works employed thousands of staff, with the Chief Mechanical Engineer or Locomotive Superintendent presiding over the operation. Although the production of a new locomotive often started with the inspirational designs of an engineer such as Bulleid, Churchward, Gresley or Stanier, the end product was the result of the efforts of many individual staff within a works. The powerhouse of the operation was the drawing office, where initial designs were transformed into working drawings for the hundreds of component parts of an engine. In the case of private locomotive builders, engines were usually built to specifications supplied by customers, whereas, in the case of railway companies, requests for particular locomotive types arose as a result of instructions from the General Manager or departments within the railway itself.

Drawing offices were often part of larger office complexes required to service the works; at Swindon, the airy space occupied by draughtsmen was on the top floor of a block full of accountants, clerks and other office staff who dealt with the ordering and specification of raw materials, managed and paid the staff, and kept accounts and records of the activities of the department. The scale of these operations varied enormously; a works such as Crewe would have employed hundreds of clerical staff in its heyday, while in 1946 a much smaller factory, the LNER plant at Inverurie in Scotland, had a complement of eleven office staff and one draughtsman.

Away from the relative calm of the works offices, the real business of building locomotives began in the 'hot' shops, where metal components were forged or cast. Larger parts such as axles, cranks and valve gear were forged using large steam hammers. Smaller items were hand-forged in blacksmiths' shops, where staff used a variety of hand tools to produce engine fittings. Smaller steam and drop hammers were gradually introduced to improve the efficiency and quality of such workshops, with other machines being used to forge small parts used in large quantities, such as nuts, bolts and rivets. The staff employed to produce castings were among the most skilled in railway factories; perhaps the most important of these were the patternmakers, whose job was to make a wooden pattern of the item to be manufactured. This was then pressed into sand contained in moulding boxes, ready for pouring with molten metal from a nearby cupola.

Most locomotive works had an iron foundry where parts such as wheels, cylinders and chimneys were cast, and a non-ferrous foundry where brass and gunmetal fittings such as whistles, safety valves and other cab fittings could be produced. In later years many works also had mechanised foundry operations where mass-produced parts such as brake blocks and track chairs could be made on a production line. At the former Caledonian Railway St Rollox Works

By modern standards the office equipment used by railway staff may seem primitive. Dorothy Boulter, the secretary to the Chief Carriage and Wagon Draughtsman at Swindon, has only a typewriter and two telephones at her disposal in this 1966 picture.

Drawing-office staff, including a very young office boy, at Swindon Works in 1891.

Making springs at Swindon in 1983: nineteenth-century working methods still in use not long before the factory closed for ever.

in Glasgow the iron foundry was turning out over 140 tons of castings a week as late as 1950.

Once castings and forgings had been produced, there was still much work to be done before parts and fittings could be assembled as part of a finished locomotive. Large numbers of staff were required in fitting and turning shops, using not only the humble file, hammer and chisel, but also more sophisticated machine tools used for all manner of specific components. Lathes, drills, milling machines and borers were just some of the many machine tools used, initially belt-driven by stationary steam engines, until the widespread introduction of electricity in the twentieth century.

The dramatic improvement in the quality and efficiency of steam locomotive boilers by 1900 was a direct result not only of better design and materials, but also of the introduction of new tooling and machinery. By 1914 most boiler shops had modern presses and rolling machines, and hydraulic riveting equipment; over forty years later much of this technology was still in use, and at Darlington Works the boiler shop was still producing

The Erecting Shop at Derby Locomotive Works in 1950.

One of the stationary steam engines used in Swindon Works to generate power around the factory. Unfortunately this wonderful piece of engineering was scrapped before preservationists could save it.

over 140 new boilers annually in the 1950s. Such workshops remained thoroughly unpleasant places to work, especially since ear protection was rudimentary, even after 1950, and at Swindon boilersmiths were known as 'fitters with their brains knocked out'.

Most works also had a wheel shop where wheels for new locomotives and tenders could be prepared, and wheels from engines in the works for

The boiler shop at the former Great Central and LNER works at Gorton in Manchester, pictured in 1950.

Special lathes were required to turn the axle journals, as well as those for turning the tyres of locomotive wheels.

overhaul could receive attention. The steel tyres used in locomotive driving wheels were heated up and then shrunk on to the wheels themselves. Wheels were then assembled on to axles using powerful presses. At Glasgow's St Rollox Works this was done by a 600-ton hydraulic wheel press. Once this process had taken place, large wheel lathes were used to ensure that the tyres were turned to the correct profile. Machines were also used to balance wheels so that they ran smoothly at high speed, and there were also other specialised machines used for quartering crank pins and grinding axle journals.

The jigsaw of component parts making up a steam or diesel locomotive was finally assembled in the works erecting shop. Engines were built over a matter of weeks, beginning with the frames, with workers adding cylinders, boilers, wheels and axles, valve gear and other fittings, cab and cab equipment, brake gear, piping, and hundreds of other components making up the final product. Although early locomotive-erecting shops were often small, dark and cramped, by 1914 many companies had replaced them with modern airy buildings with large overhead cranes capable of lifting whole locomotives, not only to wheel them, but also to move them around the workshop. At Swindon the works was transformed by the construction of the huge 'A' Erecting Shop in 1922. Completed engines were painted and varnished; although some works such as Crewe had separate paint shops, in

The 'A' Erecting Shop at Swindon in the 1950s, with various locomotives under repair.

other locations this process took place in the erecting shop itself.

Once complete, locomotives would be prepared for traffic. Most large workshops had a weighbridge where engines could be weighed and springs adjusted; when this was done, they would be rostered on to a local train enabling them to be 'run in' before being put into the locomotive fleet. At Swindon engines could also be tested on the test plant. Built by the GWR in 1904, this was, until the completion of a further plant at Rugby in 1948, the only place where steam locomotives could be run at high speed on a test bed. In the case of private builders such as North British or Vulcan Foundry, which were building many engines for export, completed locomotives were usually steamed, inspected, and then transported to nearby docks for shipping abroad.

Most workshops had powerful overhead cranes to move entire locomotives around the works easily. Here at the Brush works in Loughborough a Class 92 electric locomotive is ready to be united with its bogies.

The works of the Big Four companies did not undertake any major construction of diesel or electric locomotives until after the Second World War. The LMS had built its first main-line diesel at Derby months before nationalisation and, while various prototypes had been built, steam locomotives were still only one-third the price of new diesel or electric engines in this period. The announcement in the 1954 BR Modernisation Plan that 2,500 main-line diesel and electric locomotives were to be built to replace aging steam stock changed matters for good, and works such as Crewe, Derby, Doncaster and Swindon all built new diesel or electric

The gloom of the erecting and repair shop of the LNER's Inverurie Works, with the frames and cab of a former North British 0-6-0 in the foreground.

The erecting shop at Darlington Works, photographed by the LNER in 1946.

locomotives, using engines and electrical equipment bought in from private firms. Staff employed at the works absorbed new skills, and BR was forced to re-equip workshops to deal with the new type of motive power. Some new-build work was given to private companies such as Brush and English Electric, which had been building diesel and electric locomotives since before the Second World War, so that the new fleet that replaced steam was a mixture of BR works and privately built stock.

BR 'Britannia' 4-6-2 no. 70005 *John Milton* running at speed on the static test plant at Rugby.

While the production of new engines was the most glamorous function of railway-owned locomotive workshops, the bulk of the work carried out by the thousands of staff employed in such establishments related to the repair and maintenance of the large numbers of locomotives operated by domestic railways. Although simple repairs were carried out at engine sheds, more complicated work and routine overhauls took place in the works. Entering the factory, engines were normally stripped down, components were cleaned, inspected, and then sent to the fitting shop for attention if required, before being returned to the

Above: The front cover of a 1970s guide to the BR works at Doncaster.

Centre: New diesel construction work at Darlington in 1959.

Left: Ex-GWR 2-6-2 no. 8100 sits outside Swindon Works after overhaul in the 1950s.

erecting shop. Boilers were also removed and sent to the boiler shop for stripping and inspection before the engine was reassembled. The time spent in the works by engines for overhaul depended on the mileage they had run. At Swindon a 'light' overhaul might take a few weeks, whereas a 'heavy general' overhaul could require the engine to remain at the works for several months at least.

The heavy equipment and skilled workforce of locomotive establishments were used for more than the repair and construction of steam and diesel locomotives. Most railway-owned works were also responsible for various other parts of the working railway; the LNER works at Gorton in Manchester also maintained and repaired locomotive turntables and coaling stages in what had been the old Great Central Railway's territory. Because the GWR owned a large and important dock system after 1923, Swindon Works also became responsible for equipment such as dock gates, cranes and coal hoists. It also maintained water troughs, locomotive water cranes and stationary steam boilers and pumps, including those for the large and important pumping station that kept the Severn Tunnel clear of water. Workshops also maintained smaller outstations such as pumping houses, and some of the more complex cranes and hoists used in goods depots and sheds all over the system. Although most railways had separate permanent-way departments, many workshops had points and crossings shops where the more complicated pointwork was assembled.

Reaching the end of a tour around a locomotive works, the casual visitor would have been impressed by the scale of the place. The sheer size of the

Locomotives under repair at Darlington Works in 1959.

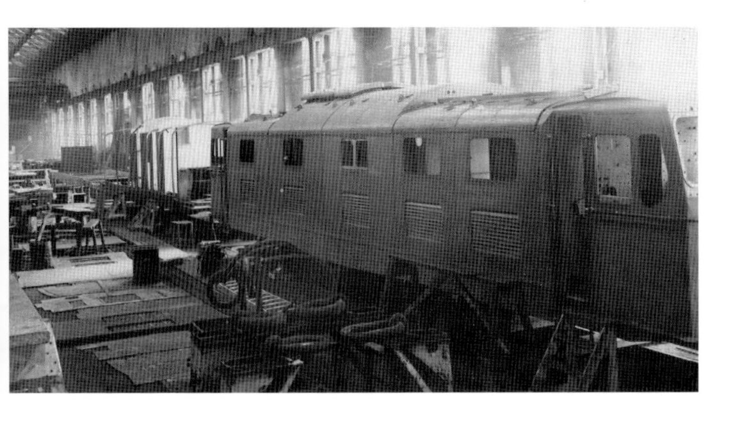

The erecting shop at Gorton showing the bodies of electric locomotives under construction in 1950.

buildings, the machinery and the engines themselves was then, and still is today, hard to comprehend. The large number of staff employed in these great enterprises also still impresses; more than seven thousand people were still working at Crewe Locomotive Works in the 1950s. Staff spent their whole lives working in the same works and it was often a family tradition to be employed by the company: in 1946, a guide to the LNER works at Inverurie in Scotland noted that it employed three generations of the Forbes family there, all named Gordon, aged respectively seventy, forty-six and twenty-one years.

A Sunday afternoon tour of Doncaster Works on 20 May 1962 and ex-LNER A4 no. 60012 *Commonwealth of Australia* is seen under repair. The engine was withdrawn on 20 August 1964 and scrapped by Motherwell Machinery & Scrap at Wishaw in May 1965.

CARRIAGES, WAGONS AND INFRASTRUCTURE

THE TASK OF BUILDING CARRIAGES, wagons and other items of equipment for railways has always proved less attractive to railway enthusiasts and historians than the building of locomotives, although the work done by railway workshops in that area was vitally important to the running of railways in Britain and overseas. While there was less heavy engineering involved in building rolling stock, the work was nevertheless very labour-intensive, illustrated by the fact that the LMS calculated that 24,000 holes might need to be drilled during the construction of one of its carriages at Wolverton Works.

Timber was the main raw material in the production of both carriages and wagons for more than a hundred years; it was usually delivered to railway works and then left outside in drying sheds to season for a long period before being brought into the sawmill to be cut to size. In later years, railway companies invested in artificial seasoning plants to speed the process. At the LNER's York Carriage Works, special drying equipment was installed in the 1930s to remove moisture from timber before its use in new rolling stock. Even with this technology it could still take between two and four hundred hours to prepare wood for use, depending on its size and type.

Before the First World War most carriages were of all-timber construction. Following a number of serious accidents in which carriages, often lit by gas, had caught fire, steel was subsequently adopted for coach underframes, and by the 1930s visitors to carriage works such as Swindon would have seen craftsmen building coaches that had wooden bodies clad in steel sheet. Eventually all-steel carriages were built. The construction and fitting out of carriages took place in large carriage-body shops, with the various component parts manufactured in separate workshops close by. Even when carriages contained far more steel, their interiors still included large amounts of timber in the form of panelling, seat backs and other fittings produced in finishing shops.

Another important workshop was the trimming shop, where staff made and repaired seat cushions, blinds and other fabric fittings, including towels,

Opposite: The size of a railway carriage wheel is apparent in this 1957 image of a carriage builder at Swindon.

Timber is sawn to a more manageable size at the LNER's Faverdale carriage works at Darlington in 1947.

rugs and blankets used in carriages and at other locations such as railway-owned hotels, refreshment rooms and ships. Most shops of this type also included a section that produced the leatherwork used in carriages, such as window straps, as well as other equipment such as cash bags and tackle used by railway-owned horses. Many women worked in these shops, carrying out

Making luggage-rack netting at Swindon Carriage Works in the 1930s.

The blacksmiths' shop at Inverurie in 1946. Apart from the rudimentary electric lighting, the equipment was broadly similar to that used by railway staff at least sixty years earlier.

much of the delicate sewing work and other tasks such as the production of luggage-rack netting.

Carriage works also usually contained separate blacksmiths' shops to produce heavier components such as buffers and couplings, although in later years some of this work was done in stamping shops, where machinery was

Carriage destination boards are being lettered in this photograph of the paint shop at Swindon Carriage & Wagon works in the 1950s.

Most carriage works had facilities to remove vermin and bugs from carriage seating and trim. This image shows what workers at Swindon called the 'Bug House'.

used to speed production. Like locomotive works, carriage works also had their own dedicated workshops equipped with wheel lathes for profiling carriage and wagon tyres, and machinery for grinding wheel journals and centres. The production and repair of metal fittings such as axleboxes, brake components, carriage bogies and vacuum pipes was usually undertaken in fitting and machine shops, where other equipment for the railway such as platform ticket machines, station barrows and trolleys, and parts for road vehicles were also manufactured. The former Caledonian Railway works at St Rollox in Glasgow also had a special shop for overhauling carriage brasswork, with an electroplating section to deal with the copper, nickel and chrome plating used in carriage toilet fittings and door handles. The so-called

LNER carriage bogies receiving attention at Inverurie before being reunited with the jacked-up teak coach bodies.

The machine shop at York Carriage Works pictured after the Second World War. Many of the larger machines are fitted with extraction systems to remove sawdust. Often workshops recycled sawdust for use in furnaces and heating.

'Utility Shop' of the GWR's carriage department at Swindon was where carpenters and cabinetmakers constructed ticket racks, station benches, desks, tables, chairs and other office furniture, signs and other equipment.

With the huge numbers of carriages in service on Britain's railways, most companies also had extensive repair facilities for their rolling stock. In 1947 the LNER carriage works at York was repairing over sixty carriages each week; at Swindon, the carriage repair shop covered an area of more than

The road-motor shop at Swindon in the 1920s. Note the large gaslight fittings in the roof.

The cover of a guide to the Litchurch Lane Works at Derby.

7 acres and could house up to 250 coaches. The GWR also built a disinfecting plant close by, used to kill vermin that had infested carriage upholstery. Most repair shops also had purpose-built cranes or jacks to lift carriage bodies on and off their bogies, speeding the repair process.

When completed, or following a heavy repair, carriages would be moved to a separate paint shop, where they would be painted, varnished and lettered. In the Victorian era a Midland Railway carriage built at Derby required twenty-six coats of primer, filling, paint and varnish in a process that might take up to three weeks. In later years, this task had been reduced to four coats and a turnround time of only four days.

Until railways began to purchase large numbers of road vehicles from established motor manufacturers, they built and maintained many of their road vehicles in carriage works, where the skills of coachbuilders were put to good use. In the nineteenth century staff were employed to build and maintain the thousands of horse-drawn drays and wagons used by goods departments. As motor lorries and omnibuses were gradually introduced, road-motor shops were adapted to cope with the new technology. Even as late as 1954 it was reported that the London Midland Region of BR had more than sixteen thousand road vehicles, maintained by the works at Wolverton.

Many of the larger works, such as Ashford, Derby, Horwich and Swindon, had their own wagon works, although there were also separate establishments, for example at Bromsgrove, as well as numerous private companies that manufactured and repaired stock for both the main-line companies and for private owners such as collieries and steelworks. The scale of this work should not be underestimated, even if it lacked the glamour of locomotive and carriage departments; in 1949 5,004 wagons received heavy repairs at Cowlairs Works, with a further 9,286 receiving light repairs.

The end product: a 12-ton wagon produced at the LNER carriage and wagon works at Faverdale in 1947.

Because of the huge variety of work undertaken in both locomotive and carriage and wagon works, staff needed to be both adaptable and skilled. Railway companies played a key role in the education of their workforce, many supporting local technical schools and colleges where apprentices could receive training in engineering. Before the Second World War, staff in both locomotive and carriage and wagon works were apprenticed in particular trades, learning 'on the job', and receiving both formal and informal training. At the end of their apprenticeship there was no guarantee that they would be employed permanently, and, although most did continue to work for the railways, in the 1920s and 1930s, when the railway industry struggled, this was not always the case. After 1945, since more staff carried out repair work than new construction, a higher proportion of skilled labour was required. This necessitated the employment of no fewer than eleven thousand apprentices at any one time, and this made British Railways one of the largest employers of apprentices in the country by the 1950s. To help matters, BR set up its own schools to train apprentices at works such as Crewe, Doncaster, Swindon and Wolverton. Working a forty-four-hour week, apprentices were, the rules noted in 1954, expected to 'submit willingly and cheerfully' to the discipline exercised over all employees.

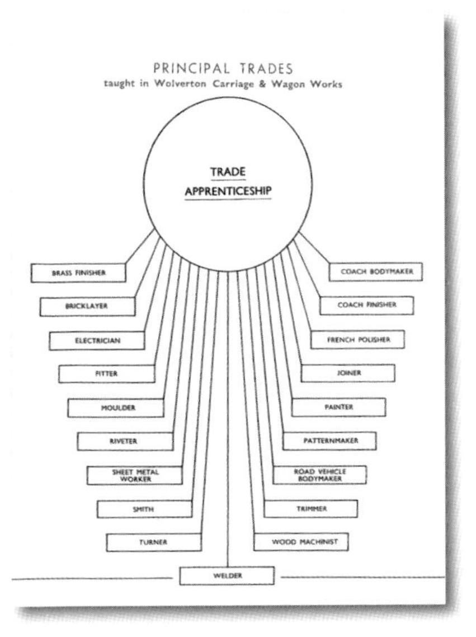

A diagram showing just how many different trades were involved in carriage and wagon manufacture and repair.

By the early twentieth century many of the best-known railway works operated by main-line companies were huge enterprises including both the locomotive and rolling-stock manufacturing and repair facilities already described. To support all this activity, factories required considerable infrastructure and services to enable them to work efficiently, and many were self-sufficient, having the tooling and expertise to produce everything they needed to build locomotives and rolling stock and many other items of equipment for the railway.

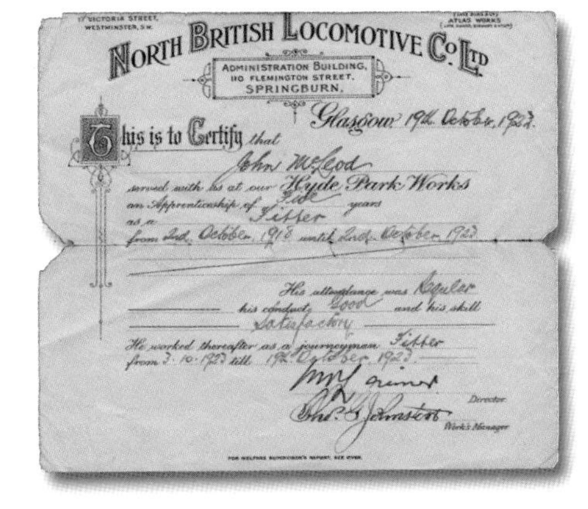
17 Victoria Street, Westminster, S.W.

Atlas Works

NORTH BRITISH LOCOMOTIVE Co. LTD

Administration Building, 110 Flemington Street, Springburn.

Glasgow 19th October 1923

This is to Certify that John McLeod served with us at our Hyde Park Works an Apprenticeship of Five years as a Fitter from 2nd October 1918 until 2nd October 1923

His attendance was Regular his conduct Good and his skill Satisfactory

He worked thereafter as a journeyman Fitter from 3.10.1923 till 19th October 1923.

Director

Works Manager

A certificate issued to John Thomas, a Journeyman Fitter employed by the North British Locomotive Company at Glasgow at the end of his five-year apprenticeship in 1920.

The facilities provided for staff in later years were much better than the primitive mess rooms originally used by workers, as this view of Swindon shows.

Maintaining an adequate supply of water was a perennial problem, since it was used in great quantities in railway works, both in industrial processes and in stationary boilers powering machinery and equipment. At Wolverton the carriage works used almost 3 million gallons of water annually, drawn from a local canal and other local sources. At Swindon the GWR was forced

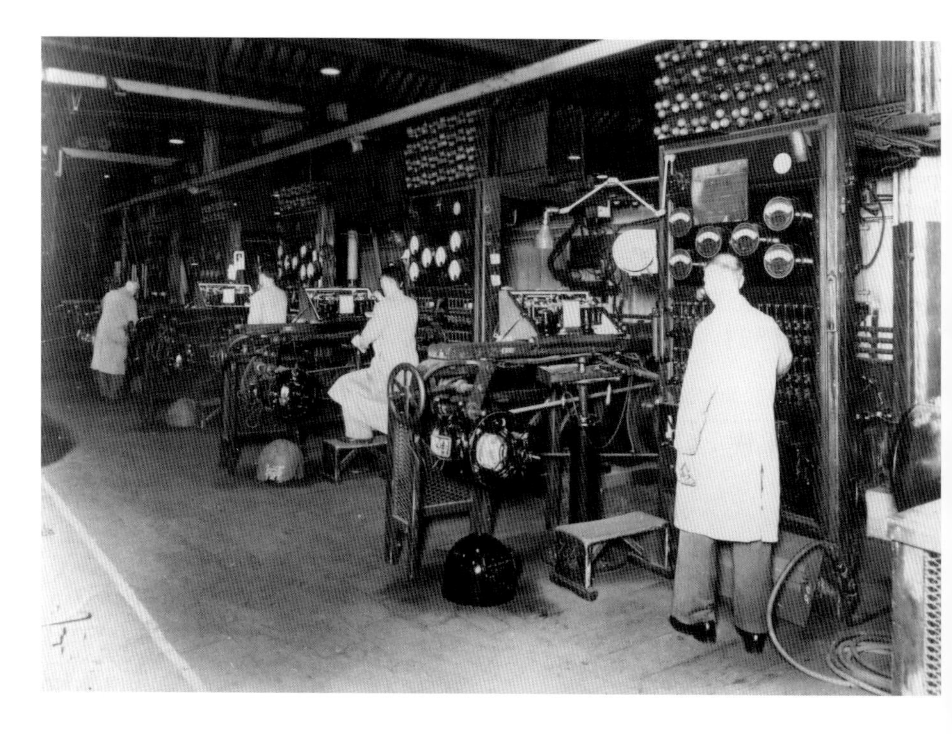

The electric shop at Wolverton Works in 1955. Here testing of train lighting took place.

to source water from a spring at Kemble some 10 miles away, and pipe it into the factory. Large gasworks were also a prominent feature since gas was initially used to provide lighting in and around workshops, as well as in furnaces and other processes. Many railway gasworks also supplied stations and railway houses close to the workshops.

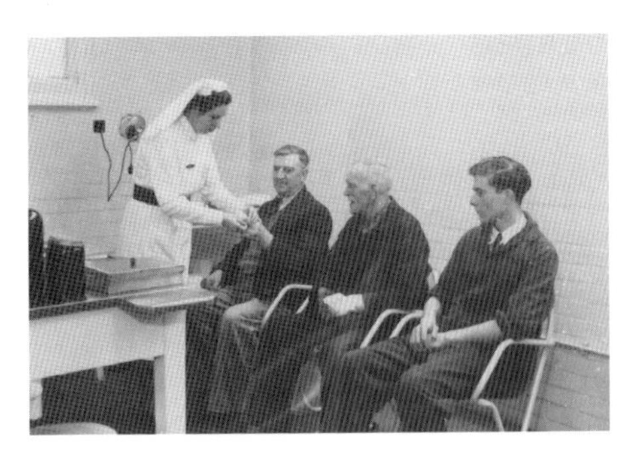

Staff at the Derby locomotive factory in the 1950s receiving medical attention from the works nurse.

By the 1930s most railway works were lit by electricity rather than gas, and in most machine shops the tangle of overhead belting used to power lathes, drills and other equipment had been superseded by the use of electric power. When the process of converting factories from gas began, most had their own power stations, but eventually electricity was obtained from municipal power companies instead. At Cowlairs Works electricity was supplied by Glasgow Corporation, the factory having its own substation built in 1935.

Steam was also required for testing and heating, washing and cooking. At Cowlairs it was supplied from eleven boilers sited around the works; in other locations, such as Swindon, power came from a large central boiler station from which steam was piped around the works site. Hydraulic power stations were also required to power machine tools, hoists and riveting equipment used in boiler shops and in both locomotive and carriage and wagon fitting shops.

Because of the large quantities of flammable wood, fabric, paint, varnish, oils and other chemicals stored and used in railway works, fire was an ever present hazard. Many larger workshops therefore had their own dedicated fire brigade. In addition, working conditions for the men and women employed in these factories were sometimes dangerous, and so railway companies had to provide first aid and medical treatment for those injured in accidents. At Ashford the Southern Railway maintained a team of 120 trained ambulance men from within the workforce, with more serious cases dealt with by a full-time attendant in a specially equipped ambulance room. Elsewhere, the St John Ambulance Brigade was used to provide first-aid cover in many railway establishments.

The Swindon Works fire engine pictured in 1984. This Dennis appliance replaced an earlier fire engine in 1942. Both are now preserved.

Although companies ran their own police forces until the creation

The futuristic nose cone of the Advanced Passenger train (APT-E), under construction at Derby in 1971.

of the British Transport Police in 1948, watchmen stood guard at the various gates and entrances within railway works to ensure that no unauthorised visitors entered, and also that works staff did not steal or remove company property. This did not prevent material being smuggled out, and many houses close to Swindon's railway works were said to have been painted in GWR colours in the age of steam.

To ensure that the raw materials bought in to build locomotives and rolling stock were of the highest quality, works such as Crewe, Derby and Swindon had their own laboratories in which they used to undertake inspection, testing and research. The Swindon works also maintained a check on the hardness of water used in GWR locomotive boilers, monitoring supplies from water cranes and tanks all over the system. The vast quantities of supplies needed to keep these huge enterprises running were kept in stores maintained by each department.

A British Railways 'Cartic', used for transporting motor vehicles, under repair at the BREL facility at Temple Mills in east London.

RAILWAY TOWNS

With the growth of the railway industry following the opening of the Stockton & Darlington Railway in 1825, new towns grew up close to the workshops and factories of both the railway companies and the locomotive manufacturers to provide accommodation for workers and their families. Some of these new communities were housed in settlements constructed by railway companies while others grew as the result of speculative development by private builders. With the exception of Crewe and Melton Constable, where works were established in rural areas with no existing settlement, most 'railway towns' as they became known, were built in or near existing villages or towns; in most cases their development dramatically transformed these places from quiet backwaters to thriving industrial centres.

In the north-east, where many of the earliest railways were promoted, the first railway town proper was established at Shildon, where the Stockton & Darlington Railway had opened its workshops in 1833. A larger development took place at Wolverton in 1838 when the London & Birmingham Railway established its workshops there. Within six years the company had built two hundred houses for the workforce, and by the early twentieth century the population of the town was itself well over seven thousand.

Two of the most famous railway towns, Crewe and Swindon, were both established in 1843. Before 1840 the workshops of the Grand Junction Railway had been at Edge Hill in Liverpool, and the site of what is now Crewe was the village of Monks Coppenhall, a rural backwater. The works at Edge Hill soon proved to be too cramped and inconveniently located for the expanding railway, and by 1842 the company engineer, Joseph Locke, had drawn up 'plans, drawings and estimates for an establishment at Crewe which shall include shops for the building and repairs of carriages and wagons as well as engines'.

By the end of 1842 thirty-two cottages had been built, and construction carried on apace until 1848, by which time the works was already employing

A house in Betley Street, Crewe, originally built for LNWR workers in the nearby railway factory.

over a thousand men, and turning out an engine a week. The operation became part of the London & North Western Railway in 1846, and over the next fifty years the town grew steadily. By 1891 the population of Crewe was 29,000, a figure that had increased to 45,000 twenty years later. The actual number of houses provided by the LNWR was, however, not great. By the 1880s there were almost five thousand houses in Crewe, but railway housing accounted for only 845 of this total.

As well as providing some housing for its staff, the railway also built a number of other facilities. One of the most important social and educational resources in Crewe was the Mechanics' Institute. As early as 1843 a library and newspaper room had been provided, and two years later the Mechanics' Institute was created, with the aim of supplying 'to the working classes of Crewe the means of instruction in Science, Literature and the Arts'. A new building to house the activities of the Institute opened in 1846, and evening classes in reading, writing, arithmetic and mechanical drawing were held there. Many social events were also run by the Institute, with dances and concerts providing workers with much-appreciated relief from the dirt and noise of the railway factory.

A further innovation was the installation of public baths in the Institute, although by the 1860s a new bath-house had been built nearby, a facility still in operation in the 1930s. The Institute itself eventually became part of what became known as the Crewe Technical Institute, but for many years provided basic engineering training for thousands of LNWR and LMS apprentices who passed through the factory.

The Great Western Railway made the decision to site its railway works at Swindon in 1840, following a visit by Isambard Kingdom Brunel and Locomotive Superintendent Daniel Gooch. The works was established on green fields some distance from the old town, close to the station and junction for the Gloucester branch. Since there was little tradition of heavy engineering in Wiltshire, the majority of the workers employed in the new factory were drawn from existing centres of railway engineering in Lancashire, Scotland and the

An Edwardian postcard view of men streaming out of the railway works at Swindon. The walls of the factory can be seen in the background.

The grand theatre of the Mechanics' Institute at Swindon, pictured before it was destroyed by fire in the 1930s.

north-east. With so little accommodation available in the town, the construction of new housing was a priority, and within a few years the railway had built three hundred cottages to Brunel's design.

The new settlement was originally called New Swindon to differentiate it from the old market town, which was some distance away. The company and the workforce co-operated to provide amenities for the families arriving in the town, and, as at Crewe, one of the first major developments was the Mechanics' Institute, an organisation that at first held events in the works itself, before constructing its own building in 1854, situated in the centre of what became known as the 'Railway Village'. The Institute also eventually came to include a library, theatre, meeting rooms and dance hall, and was managed by the workforce with some financial assistance from the GWR.

In many cases railways provided support for local hospitals and medical facilities, both as an altruistic gesture towards the workforce and their families, and also to ensure that first-aid treatment was available for staff injured in accidents inside the works. At Swindon the GWR had initially helped provide a doctor for the inhabitants of the Railway Village, but in 1847 it also supported the establishment of the Medical Fund Society, an organisation funded by weekly contributions from workers' wages, which eventually grew to include a modern hospital and dispensary, dentist's surgery, chiropodist and optician. Its headquarters also included washing, swimming and Turkish baths, essential for workmen, especially in an era when many railway houses did not have bathrooms, and families had to make do with a tin bath for washing.

From 1847 the Lancashire & Yorkshire Railway had its main locomotive works at Miles Platting in Manchester. By the 1870s, although more than 450 engines had been built there, the cramped site, surrounded by houses and other industry, was becoming unusable. The railway therefore made the drastic decision to relocate its operations to Horwich near Bolton, and new railway workshops were completed there in 1887. Horwich had relied on the cotton trade for its livelihood but had hit hard times by the time the Lancashire & Yorkshire arrived there. More than four hundred houses were already standing empty when the workshops opened, relieving the company of the need to build any for its staff. The streets close to the works were

The kitchen of a railwayman's cottage now preserved in Swindon's Railway Village.

named after famous engineers and, as at Crewe and Swindon, the town was provided with a mechanics' institute and other facilities. The coming of the works transformed the fortunes of Horwich, and within a decade of its opening the town's population had trebled.

When the directors of the South Eastern Railway announced that they were to build a 'Locomotive Establishment' at Ashford in 1847, they agreed to provide a railway village, initially of seventy-two workmen's cottages, in addition to the workshops. Named Alfred Town, and later Ashford New Town, the village grew by another sixty houses in 1850. The town also included a mechanics' institute and a school. By 1912, when electricity was introduced to the works, the new town was extended by another 126 six-roomed houses, making a total of 272 in all, and a second railway school was opened for the education of girls. Instead of belonging to a pay-related medical scheme like the one at Swindon, practically all the works staff at Ashford had contributed to a hospital fund begun in 1919 with the encouragement of the Chief Mechanical Engineer, R. E. L. Maunsell. Members of this fund voluntarily donated £1,900 towards the cost of the new Ashford hospital, which came to be a huge asset to both railway families and other residents of the town.

Ashford staff also had the use of a bowling green and tennis courts, opened following a substantial donation from the Southern Railway. There was also a cricket ground situated on railway property near the works. Most railway factories supported large numbers of sports clubs, which competed locally, regionally and nationally with other railway teams. The GWR Athletic Association at Swindon had its own purpose-built sports ground in the town, while at Gorton Works the LNER (Manchester) Athletic Association had two hundred members and ran teams for football, athletics, bowling, boxing, cricket, netball, hockey, swimming and tennis.

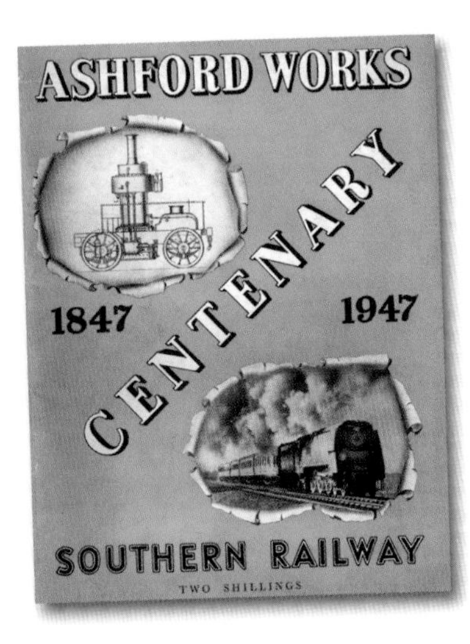

The front cover of a brochure issued to mark one hundred years of railway influence at the Kent town of Ashford.

Sport was also an important part of the social activities provided at the LNER workshops at Inverurie in Scotland. The works, opened in 1901, was modest in comparison to larger operations such as Crewe or Doncaster, with only 847 staff on the payroll in 1947. Adjacent to the factory was the Railway Institute, controlled by the grandly titled 'Inverurie Railway Recreational Association Committee'. Social gatherings such as dances, whist drives and concerts were regular features, and outside there was a bowling green, tennis courts, football pitches and children's playground. There were also 159 houses adjoining the works, originally built by the Great North of Scotland Railway for staff and their families.

Staff and patients at the GWR Medical Fund hospital in the Victorian era.

Two of the largest and most famous railway works, Derby and Doncaster, were not built on greenfield sites in rural areas, but close to existing, well-established towns. Nevertheless, these settlements soon developed their own distinctive railway communities close to the factories. At Derby the Midland Railway did not build housing for its employees, this being provided by private builders, but it did support an institute, school and orphanage. The works at Doncaster, opened in 1853 by the Great Northern, grew steadily in the nineteenth century, and the impact of more than three thousand people moving into the town was considerable. Although housing was not provided by the company, the chairman of the GNR did ask shareholders and directors to make a financial contribution towards the spiritual welfare of the workforce, part-funding the construction of St James's Church in the town, a pattern repeated in other railway towns, including Swindon.

The bowling green at Inverurie, pictured in 1947, was one of the facilities provided for staff at this LNER works.

The old roundhouse at Derby has been restored and is now used as a conference facility, part of the city's 'Railway Quarter'.

In Derby and Doncaster, the influence of the railway was perhaps not as pronounced as it was in Crewe and Swindon. These latter locations were, to a great extent, one-industry towns until the 1920s, when new industries began to provide alternative jobs for workers. Until that time employment prospects were limited and communities were extremely close-knit. This was most strikingly illustrated by the 'Trip' holiday enjoyed by GWR workers at Swindon. Every July staff were given a free ticket to allow them and their families to travel to holiday resorts for up to a week's holiday. By the First World War more than 25,000 people went 'on Trip', leaving the town largely deserted, with many traders closing for the duration themselves. In towns such as Ashford, Crewe and Swindon the railway touched all aspects of life and dominated economic, social and political activity. Underlining this influence, many railwaymen served as local councillors and mayors, and in Swindon, when 'New Swindon' and 'Old Swindon' were combined in 1901, the first mayor of the new borough was G. J. Churchward, the GWR's Locomotive Superintendent.

PLACES TO VISIT

Crewe Heritage Centre, Vernon Way, Crewe CW1 2DB. Telephone: 01270 212130. Website: www.creweheritagecentre.co.uk The centre tells the story of Crewe, its works and other parts of the town's history.

Derby College, Roundhouse Road, Pride Park, Derby DE24 8JE. Telephone: 0800 028 0289. Website: www.derby-college.ac.uk The old Midland locomotive works, now part of Derby's college, has been sensitively restored and can be viewed on organised tours.

Manchester Museum of Science and Industry, Liverpool Road, Manchester M3 4FP. Telephone: 0161 832 2244. Website: www.mosi.org.uk Includes examples of Manchester-built Beyer Peacock locomotives.

The National Railway Museum, Leeman Road, York YO26 6XJ. Telephone: 01904 621261. Website: www.nrm.org.uk The collections at York include much material about company-built railway works and their products.

Riverside Museum: Scotland's Museum of Transport and Travel, 100 Pointhouse Place, Glasgow G3 8RS. Telephone: 0141 287 2720. Website: www.glasgowlife.org.uk This award-winning museum includes displays about the North British Locomotive Company.

STEAM: Museum of the Great Western Railway, Kemble Drive, Swindon SN2 2TA. Telephone: 01793 466646. Website: www.steam-museum.org.uk The story of Swindon's railway works and the people who worked there. The Outlet Village shopping complex next door is also situated in the old workshops.

Although many works buildings have been demolished, others have been put to new uses. Part of Swindon's old railway workshops now houses the STEAM Museum, opened in 2000.

FURTHER READING

Atkins, Philip. *The Golden Age of Locomotive Building*. Atlantic Books, 1999.

Bryan, Tim. *All in a Day's Work: Life on the GWR*. OPC, 2004.

Larkin, Edgar. *An Illustrated History of British Railways Workshops*. OPC Books, 1992.

Lowe, James. *British Steam Locomotive Builders*. TEE Publishing, 1989.

Lowe, James. *Building Britain's Locomotives*. Moorland Publishing, 1979.

APPENDIX: WORKSHOPS AND MANUFACTURERS IN 1925

MAJOR BRITISH RAILWAY WORKSHOPS OPERATING IN 1925

Name	**Type**	**Railway**	**Opening Date**	**'Big Four' owner**
Ashford	L, C & W	South Eastern & Chatham	1847	SR
Brighton	L & C	London Brighton & South Coast	1852	SR
Cowlairs	L, C & W	East Glasgow & North British	1841	LNER
Crewe	L	Grand Junction & London & North Western	1843	LMS
Darlington	L	Stockton & Darlington & North Eastern	1863	LNER
Derby	L, C & W	North Midland & Midland	1840	LMS
Doncaster	L, C & W	Great Northern	1853	LNER
Eastleigh	L, C & W	London & South Western	1910	SR
Gorton	L, C & W	Great Central	1848	LNER
Horwich	L	Lancashire & Yorkshire	1887	LMS
Inverurie	L,C & W	Great North of Scotland	1902	LNER
Stratford	L	Eastern Counties & Great Eastern	1847	LNER
St Rollox	L, C & W	Caledonian	1854	LMS
Swindon	L, C & W	Great Western	1843	GWR
Wolverton	C & W	London & Birmingham & LNWR	1838	LMS
York	C & W	North Eastern	1884	LNER

INDEPENDENT RAILWAY MANUFACTURERS IN 1925

Company	Location	Type	Opening date	
Armstrong Whitworth	Newcastle	L	1847	(1)
Avonside Engine Company	Bristol	L	1837	(2)
Andrew Barclay	Kilmarnock	L	1840	
Beyer Peacock	Manchester	L	1854	
Birmingham Carriage & Wagon Company	Birmingham	L, C & W	1854	
Gloucester Carriage & Wagon	Gloucester	L, C & W	1860	
Hudswell Clarke	Leeds	L	1860	
Hunslet	Leeds	L	1864	
Kerr Stuart	Stoke-on-Trent	L	1881	(3)
Metro-Cammell	Birmingham	C & W	1863	
North British Locomotive Company	Glasgow	L	1903	(4)
Peckett & Sons	Bristol	L	1864	(5)
Robert Stephenson	Newcastle	L	1823	
Vulcan Foundry	Warrington	L	1830	(6)
Yorkshire Engine Company	Sheffield	L	1865	

(1) Originally formed as W. G. Armstrong & Co.
(2) Began life as Stothert & Slaughter & Co.
(3) Original company founded in Glasgow; moved to Stoke in 1893.
(4) NBL formed as an amalgamation of Dübs & Co, Neilson Reid and Sharp Stewart.
(5) Company founded as Fox Walker.
(6) Originally Charles Tayleur & Company.

Only the modern orange recycling box outside one of the railway houses reveals that this image of railwaymen's cottages in Swindon was taken very recently.

INDEX

Page numbers in italics refer to illustrations